NATURE
The Penguin

Barbara Todd

Established in 1907, Reed Publishing (NZ) Ltd
is New Zealand's largest book publisher, with over 300 titles in print.

For details on all these books visit our website:
www.reed.co.nz

Published by Reed Children's Books, an imprint of
Reed Publishing (NZ) Ltd, 39 Rawene Rd, Birkenhead,
Auckland. Associated companies, branches and
representatives throughout the world.

This book is copyright. Except for the purpose of fair
reviewing, no part of this publication may be reproduced
or transmitted in any form or by any means, electronic or
mechanical, including photocopying, recording, or any
information storage and retrieval system, without
permission in writing from the publisher. Infringers of
copyright render themselves liable to prosecution.

© 2001 Barbara Todd
The author asserts her moral rights in the work.
All photographs property of the author unless otherwise
credited.
© Illustrations by Richard Gunther

1 86948 874 1
First published 2001

Edited by Carolyn Lagahetau
Designed by Sharon Whitaker

Printed in New Zealand

Contents

Who, what, where?

Penguins are birds.
They live around islands in the southern oceans.
Penguins find their food in the sea and make nests
on the land.
Some penguins are endangered.

Although penguins are birds, they cannot fly. Their wing-like flippers are used for swimming.

A penguin's body is covered with two layers of feathers to help keep it warm.

5

Family tree

Penguins are closely related to small birds that are able to fly called diving petrels. There are 18 different types of penguins. Seven types live in New Zealand.

Yellow-eyed penguins are the only penguins with yellow eyes.

Yellow-eyed penguins are found only in New Zealand. Their Maori name is hoiho. Hoiho means 'noise shouter' … these penguins are very noisy!

6

Blue penguins look very tiny in the sea.

Blue penguins are the smallest penguin. They are very shy and stay in the sea until it is almost dark. Blue penguins live in New Zealand and Australia.

White-flippered blue penguins only live in New Zealand. They nest around Banks Peninsula in the South Island.

B. Bell, DoC

There are four crested penguins that nest on islands in New Zealand waters. They are the Rockhopper, Erect-crested, Fiordland crested and Snares crested penguins.

Lou Sanson

Above: Rockhopper penguin

J.L. Kendrick, DoC

Left: Erect-crested penguins

Lou Sanson

Left: Snares crested penguin
Below: Fiordland crested penguin

The Fiordland crested penguin is very rare. It lives on the west coast of New Zealand's South Island. It nests in small caves, under large rocks or among tree roots.

Good looks

The body shape of each type of penguin looks almost alike. Some penguins have different colours on their head or chest. Penguins come in lots of different sizes.

All penguins have a white front and a black back.

Lou Sanson

All crested penguins have yellow feathers on their heads.

King penguins have yellow and orange markings on their head and chest.

Once a year, penguins get a new coat.
Their old feathers fall out and they look
very funny. They cannot swim in the sea
and hunt for food until their new feather
coat grows back.

Emperor penguins are the tallest penguin. They are up to 1.3 metres tall – about the same size as a 7 to 9 year old child!

Blue penguins are the smallest penguin. They are only 40 to 45 centimetres tall – about the same size as a newborn baby.

Dinnertime!

Penguins catch their food in the sea.
When they chase fish, they look like they are flying under water.

Penguins eat small
fish and squid.

Sometimes they
eat tiny sea
animals called
krill.

Sounds

Penguins are very noisy!
They call to each other with loud sounds.
Sometimes they make sounds almost like a donkey!

Each penguin has a voice that is different from all other penguins.

R. Morris, DoC

15

Home, sweet home

Penguins live in different places.
During the day they hunt for food out at sea.
They come ashore at night to sleep.

Some penguins live where there is snow and ice.

Nathalie Patenaude

Some penguins live where it is rocky and hot.

Some penguins live where it is grassy and green.

17

Eggs

Female penguins lay their eggs once a year.
Sometimes the father keeps the eggs warm while the mother looks for food.

B. Ahern, DoC

Some penguins lay two eggs.
The parents take turns guarding the eggs while each goes to sea to search for food.

Emperor and King penguins lay one egg.
It takes 2 to 3 months before the chick begins to hatch.

B. Ahern, DoC

Emperor and King penguins keep their egg warm by placing it on their feet and covering it with their warm, feathered stomach.

Young ones

New penguin chicks are looked after by both parents.
One parent protects the chick while the other brings food.

Newborn Emperor and King penguin chicks live on their parent's feet and stay warm under their feathers.

R. Morris, DoC

Some Yellow-eyed penguins raise two chicks. The chick's bodies are covered with soft, downy feathers. When they get older, their soft feathers will be replaced with water-proof feathers, just like their parents'.

Yellow-eyed penguin chicks hatch in one day.

K. Westerskov, Hedgehog House, NZ

It takes an Emperor penguin chick 2 to 3 days to break out of its shell.

19

Danger! Beware!

Penguins face danger at sea and on land.
Sometimes penguins are eaten by killer whales in the sea. Leopard seals and large sea lions also eat penguins, both at sea and on land.

These Yellow-eyed penguins have just arrived home.
They are looking around to make sure there is no danger.
The penguins are hoping the leopard seal will not see them as they make their way to their nest in the bushes.

Dogs, cats, stoats and ferrets eat penguin eggs and young penguin chicks.

Sea lion

Skuas are fierce birds who will steal and eat penguin eggs and attack young penguin chicks.

Killer whale

Help save us

Penguins need our help.
They need sheltered places on land
to rest during the night.
They need a clean ocean where
they can find food to eat.

People from the Yellow-eyed Penguin Trust are planting bushes to provide safer, sheltered nesting sites for the penguins and their young chicks.

We can help penguins by not throwing plastic and other garbage onto the beach.

A. Tennyson, DoC

Penguins can get trapped in fishing nets or other types of garbage that has been thrown in the ocean or on the beach.

23

Index

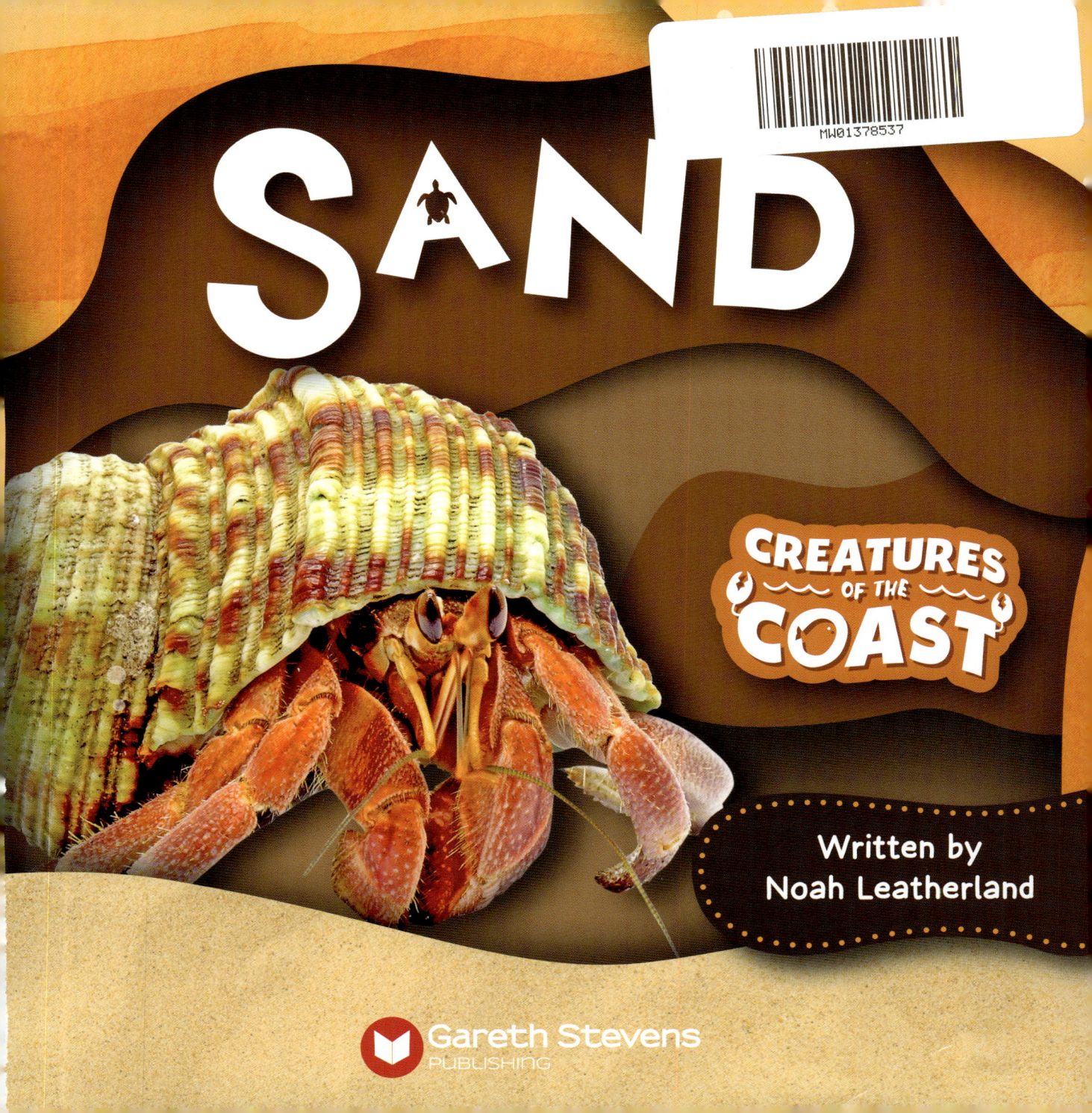

SAND

CREATURES OF THE COAST

Written by
Noah Leatherland

Gareth Stevens
PUBLISHING

Please visit our website, www.garethstevens.com. For a free color catalog of all our high-quality books, call toll free 1-800-542-2595 or fax 1-877-542-2596.

Published in 2025 by
Gareth Stevens Publishing
2544 Clinton St.
Buffalo, NY 14224

Written by:
Noah Leatherland

Edited by:
Rebecca Phillips-Bartlett

Designed by:
Ker Ker Lee

Cataloging-in-Publication Data

Names: Leatherland, Noah, 1999-.
Title: Sand / Noah Leatherland.
Description: New York : Gareth Stevens Publishing, 2025. | Series: Creatures of the coast | Includes glossary and index.
Identifiers: ISBN 9781538294680 (pbk.) | ISBN 9781538294697 (library bound) | ISBN 9781538294703 (ebook)
Subjects: LCSH: Seashore animals--Juvenile literature. | Beaches--Juvenile literature. | Sand--Juvenile literature.
Classification: LCC QL122.2 L438 2025 | DDC 553.6'22--dc23

© 2023 Booklife Publishing

This edition is published by arrangement with Booklife Publishing

All rights reserved. No part of this book may be reproduced in any form without permission in writing from the publisher, except by a reviewer.

Printed in the United States of America

CPSIA compliance information: Batch #CSGS25: For further information contact Gareth Stevens at 1-800-542-2595.

Find us on 📘 📷

PHOTO CREDITS: All images courtesy of Shutterstock. With thanks to Getty Images, Thinkstock Photo and iStockphoto.

Recurring images: Alexander_Evgenyevich, Baranovska Oksana, holaillustrations, Mia S PARK, Nesterova. Cover – Mark Brandon, Tonographer. 2–3 – Julian Wiskemanni. 4–5 – Stefan Neumann, Zack Frank. 6–7 – haveseen, Seth Yarkony. 8–9 – kzww, Matt Filosa. 10–11 – Torsten Pursche, Makarenko Igor, Alfmaler. 12–13 –David R Butler, Lavendulan. 14–15 – Henrik Larsson, Macronatura.es, zdenek_macat, Danilashchyk Olena. 16–17 – Tony Northrup, Kalaeva, zulkamalober. 18–19 – Christopher Seufert, Stewart Kirk, pisanstock. 20–21 – Mike Korostelev, Patrick Messier, SL-Photography, mollyw. 22–23 – ThomBal, Markos Loizou.

CONTENTS

Words that look like this can be found in the glossary on page 24.

IN THE SAND

If you visit the coast, you might find a sandy beach. A lot of amazing creatures make their homes in the sand. The next time you are at the coast, these creatures might be there too!

The coast is where the land meets the sea.

Some creatures spend their whole lives in the sand. Others live out at sea and only come onto the sand at certain times of the year. Keep an eye out for them!

HERMIT CRABS

HERMIT CRABS ARE OMNIVORES. THEY EAT ANYTHING THEY CAN FIND.

Did you know that hermit crabs are not born with their shells? They search through the sand for a shell to make their home. Hermit crabs might even fight each other over a shell.

Hermit crabs test shells to make sure they can comfortably fit inside. Hermit crabs need to find bigger shells as they grow. They make their homes in many shells throughout their lives.

LUGWORMS

Have you ever come across curly piles of sand on the beach? These are made by lugworms. Lugworms make <u>burrows</u> to live in by swallowing sand and pooping it out!

Lugworms can be black, brown, pink, or green. They can grow to around 16 inches (40.6 cm) long. Coastal birds hunt lugworms by pulling them out of their underground burrows.

SEA LIONS

Sea lions spend a lot of time at sea, but sometimes they gather on beaches in large groups. They climb onto the sand when they are ready to have babies.

SEA LIONS ARE MAMMALS.

MANE

Sea lions use their flippers to crawl across the sand to find a place to rest. Sea lions got their name because they are large, they roar, and some have manes just like lions do.

11

SAND DOLLARS

SOFT, HAIRY SPINES

Sand dollars are flat creatures. They have lots of soft spines and hairs covering their bodies. Sand dollars use these spines to move, eat, and bury themselves in the sand.

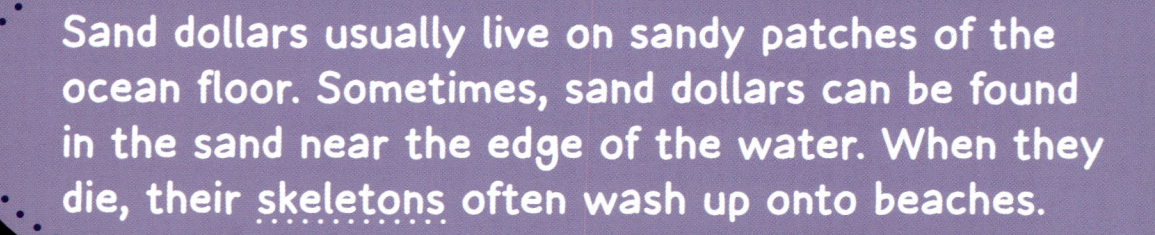

Sand dollars usually live on sandy patches of the ocean floor. Sometimes, sand dollars can be found in the sand near the edge of the water. When they die, their skeletons often wash up onto beaches.

Some stories say that sand dollars are coins dropped by mermaids!

BUGS OF THE BEACH

KELP FLY

ROVE BEETLE

Many bugs live along the coast. The moist environment of the seaside makes a perfect home for bugs such as rove beetles. Kelp flies eat slimy seaweed found on beaches.

Some bugs have <u>adapted</u> to life in the sand. Beach wolf spiders are light brown and have spots on their bodies. This works as <u>camouflage</u> that helps them blend into the sand.

BEACH WOLF SPIDER

15

SEA TURTLES

SEA TURTLE EGGS

Sea turtles spend most of their lives in the ocean, but their lives begin in the sand. Mother turtles crawl onto the beach to bury their eggs in the sand.

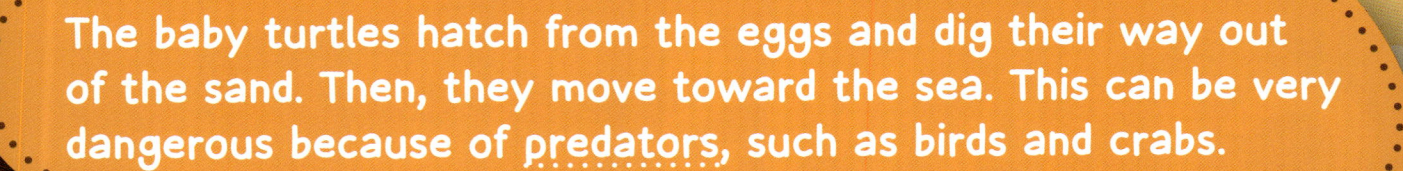

The baby turtles hatch from the eggs and dig their way out of the sand. Then, they move toward the sea. This can be very dangerous because of predators, such as birds and crabs.

CLAMS

Clams are soft creatures that live inside their shells. Some clams can be found in burrows on sandy beaches. Burying themselves in the sand protects them from predators, such as seabirds.

Clams open their shells to filter feed.

Clams eat by filter feeding. This means they take in water from around them and eat tiny pieces of food found in the water. Then the water is pushed back out.

PENGUINS

People often think of penguins living in Antarctica, but many types of penguins live on sandy beaches. African penguins live on sandy beaches in groups called colonies.

In their colonies, African penguins groom each other and work together to make burrows in the sand. These burrows are safe places to lay their eggs. African penguins are very noisy birds. They make sounds like donkeys!

CREATURES OF THE COAST

Sandy coasts are important to many creatures all around the world. Some animals spend their whole lives in the sand. Other creatures come to the beach to have babies and raise their young.

From little bugs to massive sea lions, all sorts of creatures can be seen on the sand. The next time you go to the coast, see what animals you can spot!

GLOSSARY

adapted	changed over time to suit the environment
burrows	homes made by animals by digging into the ground
camouflage	to blend into the surroundings or background
groom	to brush and clean an animal's coat or fur
mammals	animals that are warm-blooded, have backbones, and produce milk
moist	slightly wet
omnivores	animals that eat both plants and other animals
predators	animals that hunt other animals for food
skeletons	the frameworks of bones supporting the body

INDEX